EQAO Grade 6 Math Test Prep

—— Teacher Guide ——

Written by Ruth Solski

RUTH SOLSKI was an educator for 30 years. She has written many educational resources and is the founder of S&S Learning Materials. As a writer, her main goal is to provide teachers with a useful tool that they can implement in their classrooms to bring the joy of learning to children.

Published in Canada by:
On The Mark Press
15 Dairy Avenue, Napanee, Ontario, K7R 1M4
www.onthemarkpress.com

Funded by the
Government
of Canada

Canadä

Table of Contents

About This Book

This book was created to help Grade 6 students prepare for the EQAO Mathematics Assessment test. The 10 practice tests were designed to be similar to the actual test the students will be taking. The questions are either multiple choice or open response so that students can get familiar with the question/answer format.

The first 5 tests each feature a different key math skill for targeted practice: Number Sense and Numeration, Patterning and Algebra, Measurement, Geometry and Spatial Sense, and Data Management and Probability. Students struggling in any of these areas will benefit from this skill-specific practice. The following 5 tests feature mixed math skills with questions that are a combination of all the math skills.

There is no particular sequence to the tests. They can be used in whatever order you choose to fit your students needs.

SSG115   ISBN: 9781487704049 © On The Mark Press

Number Sense and Numeration

1. A fresh shipment of mangoes just arrived at the market. Each mango cost 90¢. How many mangoes can be bought for $10.80?

   ○ 9

   ○ 20

   ○ 11

   ○ 12

2. Which of the following represents the greater number?

   ○ 0.24

   ○ 2.4

   ○ 1.24

   ○ 2.04

3. Which of the numbers below comes between 85% and 95%?

   ○ 0.15

   ○ 0.61

   ○ 0.89

   ○ .98

4. What is 5% of $250,000?

   ○ $12 500

   ○ $125

   ○ $125 000

   ○ $12.50

5. Which of the numbers below expresses 6 ones and 93 hundredths?

   ○ 6.93

   ○ 6.193

   ○ .693

   ○ 6.930

Number Sense and Numeration

6. What is the cost of 24 textbooks at $35.00 each?

   ○ $940

   ○ $740

   ○ $640

   ○ $840

7. Which of the percentages listed below is equal to ½ ?

   ○ 33%

   ○ 50%

   ○ 90%

   ○ 60%

8. What is the correct way to write the number 80 185 in words?

   ○ eight thousand eighty-five

   ○ eight hundred thousand one eighty-five

   ○ eighty thousand one hundred eighty-five

   ○ eighty thousand eighty-five

9. I am a number. Subtract me from 25; divide by 5; the result is 3.

   What number am I?

   ○ 5

   ○ 15

   ○ 20

   ○ 10

10. Which of the following is the expanded numeral for 842 071?

    ○ 8 + 4 + 2 + 0 + 7 + 1

    ○ (800 × 1 000) + (42 × 1 000) + 71

    ○ (8 × 10 000) + (40 × 10 000) + 271

    ○ (800 × 1 000) + (4 × 10 000) + (2 × 1000) + (7 × 100) + 1

SSG115   ISBN: 9781487704049 © On The Mark Press

Number Sense and Numeration

11. A group of 25 tourists are taking a trip to the art museum for a guided tour. Each tourist pays the following costs for the trip:

TRANSPORTATION: $14.85

MUSEUM ADMISSION: $6.35

LUNCH: $10.70

TOUR GUIDE FEE: $2.40

Round the costs to the nearest dollar and use them to estimate the total cost for the 25 tourists.

Show your work.

The estimated total cost for all 25 tourists is _____ .

Number Sense and Numeration

12. A class of 50 college students took a science test — 35 students passed the test; 15 students failed the test.

Fill in the chart below to express the following:

- ratio of students who passed the test to the number of students who took the test

- fraction of students who passed the test (reduce to a simple fraction)

- percent of students who passed the test

- percent of students who failed the test

Ratio of Students Who Passed	Fraction of Students Who Passed	Percent of Students Who Passed	Percent of Students Who Failed

Show your work.

SSG115   ISBN: 9781487704049 © On The Mark Press

Number Sense and Numeration

13. Blaine's Department Store did a survey of its customers to see what percentage of them shopped online. The results were 40% of their customers shopped online.

    Which of the following numbers is equivalent to 40%?

    ○ 4.0

    ○ $\frac{1}{4}$

    ○ $\frac{4}{10}$

    ○ 0.04

14. A farmer sold 93 lambs. He sold 41 of them at $280 each; the rest he sold at $225 each. How much money did he receive for the lambs?

    ○ $22 180

    ○ $22 480

    ○ $23 180

    ○ $23 170

15. Zafar is training for a 10k marathon. It takes him 18 minutes to walk 1 kilometre. At that rate, how long will it take him to walk 10 kilometres?

    ○ 1 hour

    ○ 3 hours

    ○ 2 hours

    ○ 5 hours

16. The dance teacher needs to arrange 96 dancers in rows. Each row must have an equal number of dancers. Which of the following could be the method the dance teacher uses to arrange the dancers?

    ○ 7 rows of 11

    ○ 12 rows of 14

    ○ 8 rows of 9

    ○ 8 rows of 12

Number Sense and Numeration

17. Rachel and her brother ordered a pizza. When it came it was cut into 8 slices. Rachel ate ¼ of the pizza. Then her brother ate ½ of what was left.

How many slices were left for Rachel's mom to eat?

Show your work.

Rachel's mom ate _____ slices.

SSG115   ISBN: 9781487704049 © On The Mark Press

Number Sense and Numeration

18. The Martinez family lives in Windsor. They rented a van so that they could travel to Quebec City for the Winter Carnival festivities. The left at 8:00 a.m. on a Saturday morning.

Below is the distances between the cities that they travelled through on their way.

WINDSOR to **TORONTO:**. . . . . 370 km

TORONTO to **KINGSTON:** . . . . 260 km

KINGSTON to **MONTREAL:**. . . . . 290 km

MONTREAL to **TROIS-RIVIERES:** . . . . . . 181 km

TROIS-RIVIERES to **QUEBEC CITY:** . . . . . 169 km

a. What is the total distance the Martinez family travelled?

b. At a speed of 110 km per hour, how long did it take them to reach their destination without stopping?

c. Approximately what time did they arrive in Quebec City?

Show your work.

a. They travelled _____ **km.**

b. It took them _____ **to get there.**

c. They arrived at approximately _____ **.**

Patterning and Algebra

1. Which rule describes this pattern?

   2, 6, 18, 54

   ○ Start with 2 and multiply by 2 to find the next term.

   ○ Start with 2 and add 12 to find the next term.

   ○ Start with 2 and multiply by 3 to find the next term.

   ○ Start with 2 and divide by 3 to find the next term.

2. A pattern that increases when the same amount is added to each term is represented in the table below.

   PATTERN TABLE

Term Number	Term Value
1	15
2	22
3	29
4	36
5	43

   Which of the following is the term number when the term value is 64?

   ○ 8

   ○ 7

   ○ 21

   ○ 50

3. What are the four missing numbers in the third square?

2	5	4	10		8	20
3	6	6	12	?	12	24

   ○ | 5 | 11 |
     |---|----|
     | 7 | 13 |

   ○ | 11 | 9 |
     |----|---|
     | 12 | 24 |

   ○ | 6 | 15 |
     |---|----|
     | 9 | 18 |

   ○ | 6 | 12 |
     |---|----|
     | 9 | 20 |

SSG115   ISBN: 9781487704049 © On The Mark Press

Patterning and Algebra

4. What is the value of Z in the following equation?

$$Z + Z - 9 = 5$$

Show your work.

The value of Z is _____.

---

5. What is the value of X in the following equation?

$$\frac{x}{3} + \frac{x}{3} = 1\tfrac{1}{3}$$

Show your work.

The value of x is _____ .

Patterning and Algebra

6. Use the graph below to plot the following coordinates:

(X,Y): (3,1); (6,2); (9,3); (12,4)

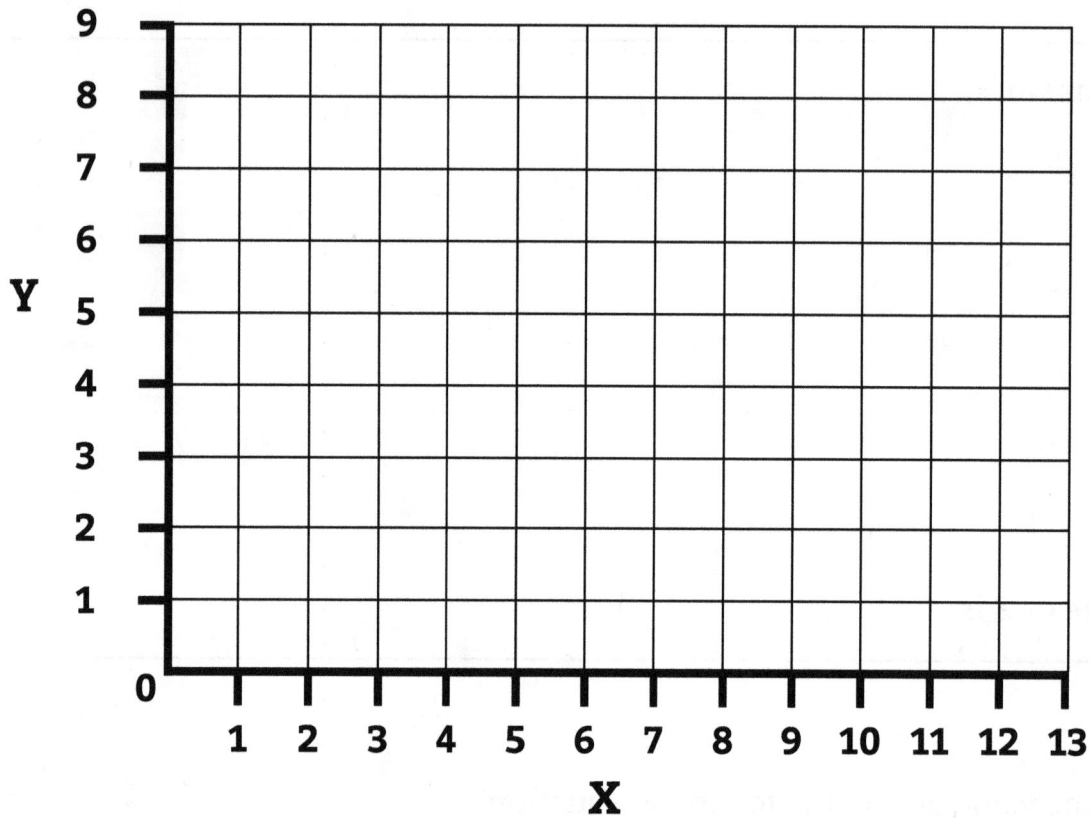

What should the next coordinate be?

Justify your thinking.

The next coordinate is _____ .

SSG115 ISBN: 9781487704049 © On The Mark Press

Patterning and Algebra

7. Emily is picking strawberries to make some extra money. On Day 1 she only picks 3 small baskets of berries. She asks a friend to help her. Each day after that they pick double the amount of baskets as the day before.

   On what day will Emily and her friend pick exactly 48 baskets?

   ○ Day 3

   ○ Day 6

   ○ Day 5

   ○ Day 4

8. Examine the input–output table shown below.

Input	Output
3	7
5	13
6	16
7	19

   Which of these rules describes the data?

   ○ Multiply by 2 and add 1.

   ○ Multiply by 3 and subtract 2.

   ○ Multiply by 3 and subtract 1.

   ○ Multiply by 2 and add 3.

9. A repeating pattern is shown below.

   ▲►▼◄▲►▼◄

   What is the 16th figure in the pattern?

   ○ ▲

   ○ ▼

   ○ ◄

   ○ ►

Patterning and Algebra

10. Look at the graph below. See if you can find the pattern. Then solve the missing numbers.

Record your answers in the spaces below the graph.

1	1	1	1	1	1	1	1	1	1
1	2	3	4	5	6	7	8	9	**H**
1	3	6	10	15	21	28	36	45	55
1	4	10	20	35	56	84	**C**	165	220
1	5	15	35	70	126	210	330	495	715
1	6	21	56	126	252	462	792	**E**	2002
1	7	28	**A**	210	462	**B**	1716	3003	**I**
1	8	36	120	330	792	1716	**D**	**F**	11440
1	9	45	165	495	1287	3003	6435	12870	**J**
1	10	55	220	715	2002	5005	11440	**G**	48620

A _____   F _____

B _____   G _____

C _____   H _____

D _____   I _____

E _____   J _____

SSG115   ISBN: 9781487704049 © On The Mark Press

Patterning and Algebra

11. A survey of the 36 students in Mrs. Zupan's sixth grade class showed that her class preferred the following hotdog toppings. Each student was asked to pick only one of the three toppings as their favourite.

Toppings	Relish	Ketchup	Mustard
Number of Students	12	18	6

If Mrs. Zupan's class is representative of sixth grade hotdog topping preferences in the school, what would you expect the results for Mr. Gordon's sixth grade class of 30 students to be?

Show your work.

Fill in the following chart with your answers for Mr. Gordon's sixth grade class.

Toppings	Relish	Ketchup	Mustard
Number of Students			

Patterning and Algebra

12. The following pattern increases by following this rule: multiply the previous term by 4 and subtract 3.

    6, 21, 81, 321, . . .

    What is the next term in the sequence?

    ○ 1 248

    ○ 1 281

    ○ 988

    ○ 1 254

13. The city park is planning a new, large flower garden. The design includes 5 rows of flowers. The first row will have 72 flowers. The second row will have 88 flowers. The third row will have 104 flowers. Each row after this continues to increase by the same number. How many total flowers will be planted in the new garden?

    ○ 480

    ○ 540

    ○ 520

    ○ 560

14. What value when placed in the box will make the following number sentence true?

    $8 \times 12 - \boxed{} = 45 + 45$

    ○ 5

    ○ 7

    ○ 8

    ○ 6

15. Look at the repeating pattern below.

    AEECCCZ   AEECCCZ

    If the pattern continues, what will the 73rd letter be?

    ○ A

    ○ Z

    ○ E

    ○ C

SSG115   ISBN: 9781487704049 © On The Mark Press

Measurement

1. Jamal ran a race that was 3.5 kilometres. How many metres did he run?

   ○ 35

   ○ 350

   ○ 3.5

   ○ 3500

2. Which of the number sentences below show the correct way to calculate the volume of the box below.

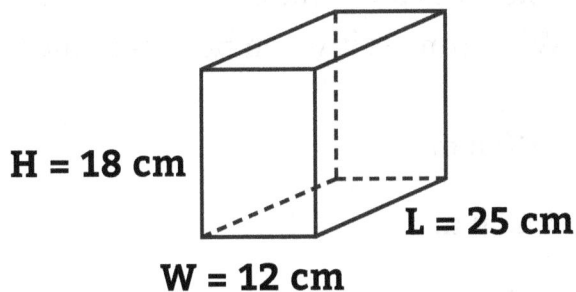

   H = 18 cm
   L = 25 cm
   W = 12 cm

   ○ 12 cm × 18 cm = 216 cm²

   ○ 12 cm × 25 cm × 18 cm = 5400 cm³

   ○ 18 cm × 25 cm = 450 cm²

   ○ 12 cm + 18 cm + 25 cm = 56 cm³

3. Pavitra has 4 items in front of her. The mass of each item is recorded below.

   ITEM 1: . . . 2000 mg

   ITEM 2: . . . 500 g

   ITEM 3: . . . 2 kg

   ITEM 4: . . . 650 g

   Which item has the greatest mass?

   ○ Item 1

   ○ Item 2

   ○ Item 3

   ○ Item 4

4. Which of the following is the most appropriate unit of measurement to describe the area of the floor of the train station?

   ○ cm²

   ○ km²

   ○ m²

   ○ m³

Measurement

5. The Lazarri family took a drive through the countryside. They drove 420 km without stopping. It took them 6 hours to get to their destination.

    At what speed were they travelling?

    ○ 42 km per hour

    ○ 60 km per hour

    ○ 70 km per hour

    ○ 80 km per hour

6. Which of the following number combinations shows the area and perimeter of the rectangle shown below.

    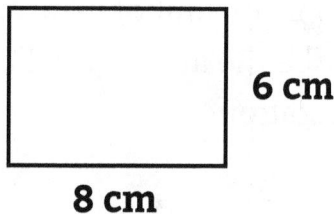

    6 cm

    8 cm

    ○ area: 28 cm   perimeter: 48 cm²

    ○ area: 48 cm²   perimeter: 28 cm

    ○ area: 28 cm²   perimeter: 14 cm

    ○ area: 14 cm²   perimeter: 48 cm

7. Asha can hit 25 tennis balls with her tennis racquet in one minute. At that same rate, how many tennis balls should she be able to hit in one hour?

    ○ 1500 tennis balls

    ○ 750 tennis balls

    ○ 250 tennis balls

    ○ 1200 tennis balls

8. Logan gets on the bus at 4:15 p.m. It takes him 50 minutes to get home. What time is it when he gets home?

    ○ 5:00 p.m.

    ○ 4:45 p.m.

    ○ 5:05 p.m.

    ○ 5:10 p.m.

SSG115   ISBN: 9781487704049 © On The Mark Press

Measurement

9. What is the area of the triangle shown below?

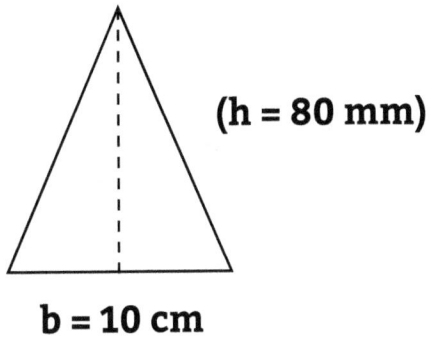

(h = 80 mm)

b = 10 cm

Write a number sentence to show your answer.

Show your work.

The area of the triangle is: _____

Measurement

10. What is the volume of the rectangular prism below?

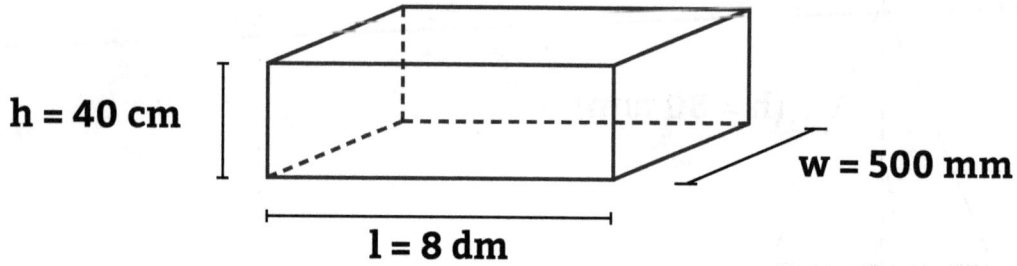

h = 40 cm

w = 500 mm

l = 8 dm

Write a number sentence to show your answer.

Show your work.

The volume of the rectangular prism is: _____

SSG115   ISBN: 9781487704049 © On The Mark Press

Measurement

11. A rectangle has an area of 162 cm². The length is 18 cm. Which of the following is the width?

   ○ 9 cm

   ○ 18 cm

   ○ 16 cm

   ○ 7 cm

12. There is 21 litres of water to share equally between 7 horses. How many millilitres does each horse get to drink?

   ○ 3 millilitres

   ○ 30 millilitres

   ○ 154 millilitres

   ○ 3000 millilitres

13. Which of the following is the equivalent to 5 m²?

   ○ 50 000 cm²

   ○ 5000 cm²

   ○ 500 cm²

   ○ 50 cm²

14. Aaron is taking a hike to the waterfall. It's 1:30 p.m. and he just stopped for lunch. Aaron has been hiking for 2 hrs. and 20 min. What time did Aaron start his hike?

   ○ 11:30 p.m.

   ○ 11:10 a.m.

   ○ 11:00 a.m.

   ○ 11:10 p.m.

Measurement

15. Mrs. Baxter has the following amount of money in her purse:

<u>**COINS**</u>	<u>**BILLS**</u>
5 **PENNIES**	2 **$5 BILLS**
3 **NICKELS**	1 **$10 BILL**
8 **DIMES**	2 **$20 BILLS**
7 **QUARTERS**	
6 **ONE-DOLLAR COINS**	
5 **TWO-DOLLAR COINS**	

Answer the following questions using the information above.

a. What is the smallest combination of bills and coins in Mrs. Baxter's purse that would make a total of $36.40?

b. What is the smallest combination of bills and coins in Mrs. Baxter's purse that would make a total of $14.68?

Show your work.

List combinations of bills and coins here:

a. _____

_____

b. _____

_____

SSG115   ISBN: 9781487704049 © On The Mark Press

Measurement

16. Shawna is helping her mom cook a holiday dinner. The turkey they are cooking weighs 8.5 kilograms. It takes 20 minutes to cook 500 grams of this turkey. Shawna is in charge of taking the turkey out when it is done.

How many minutes will it take to cook the whole turkey?

Approximately how many hours will it take?

Show your work.

It will take _____ minutes to cook the turkey.

It will take approximately _____ hours to cook the turkey.

Geometry and Spatial Sense

1. What kind of triangle is shown below?

   ○ equilateral triangle

   ○ isosceles triangle

   ○ scalene triangle

   ○ right triangle

2. How many lines of symmetry can be drawn in the figure below?

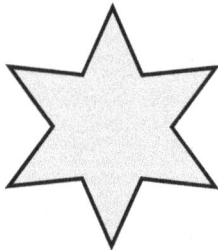

   ○ 6 lines of symmetry

   ○ 4 lines of symmetry

   ○ 8 lines of symmetry

   ○ 3 lines of symmetry

3. Nikita is building a birdhouse. She wants the roof of her birdhouse to be at an angle more than 90° but less than 120°.

   Using a protractor, which angle below could be used for the roof?

   ○

   ○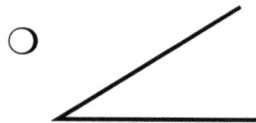

   ○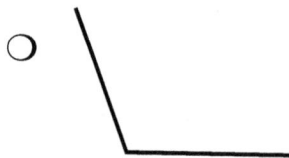

   ○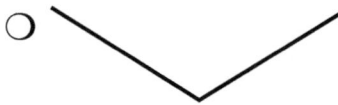

SSG115   ISBN: 9781487704049 © On The Mark Press

Geometry and Spatial Sense

4. Which figure below, when unfolded, resembles this net?

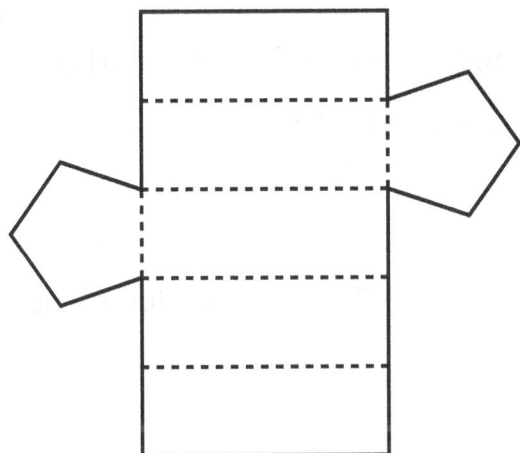

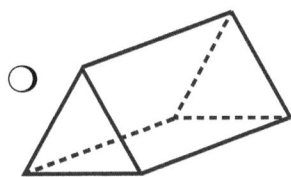

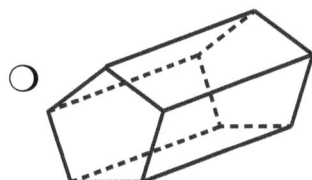

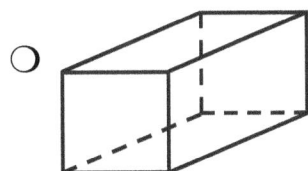

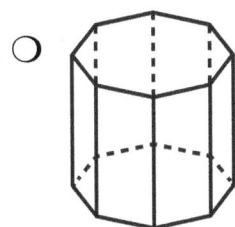

5. How many vertices does an octagon have?

   ○ 5 vertices

   ○ 6 vertices

   ○ 8 vertices

   ○ 4 vertices

6. What shape is the face of a cylinder?

   ○ rectangle

   ○ circle

   ○ square

   ○ triangle

7. What kind of angle is a 125° angle?

   ○ acute

   ○ right

   ○ reflex

   ○ obtuse

Geometry and Spatial Sense

8. Using a protractor and a ruler, draw a figure in the box below following these instructions:

   STEP 1: At point A, draw a straight line 6 cm long and mark the other end B.

   STEP 2: Using a protractor, draw a 90° angle at both ends A and B.

   STEP 3: At point A extend the right angle line 4 cm. Label that point D. Do the same at B and call that point C.

   STEP 4: Join C and D, making sure that the length of CD is the same length as line AB – 6 cm.

   STEP 5: Name the figure you have just drawn.

A•

The shape I have drawn is a _____ .

SSG115   ISBN: 9781487704049 © On The Mark Press

Geometry and Spatial Sense

9. Using the isometric dots, recreate the image below continuing from line AB. You will need a ruler to draw the lines and a protractor to measure the angles.

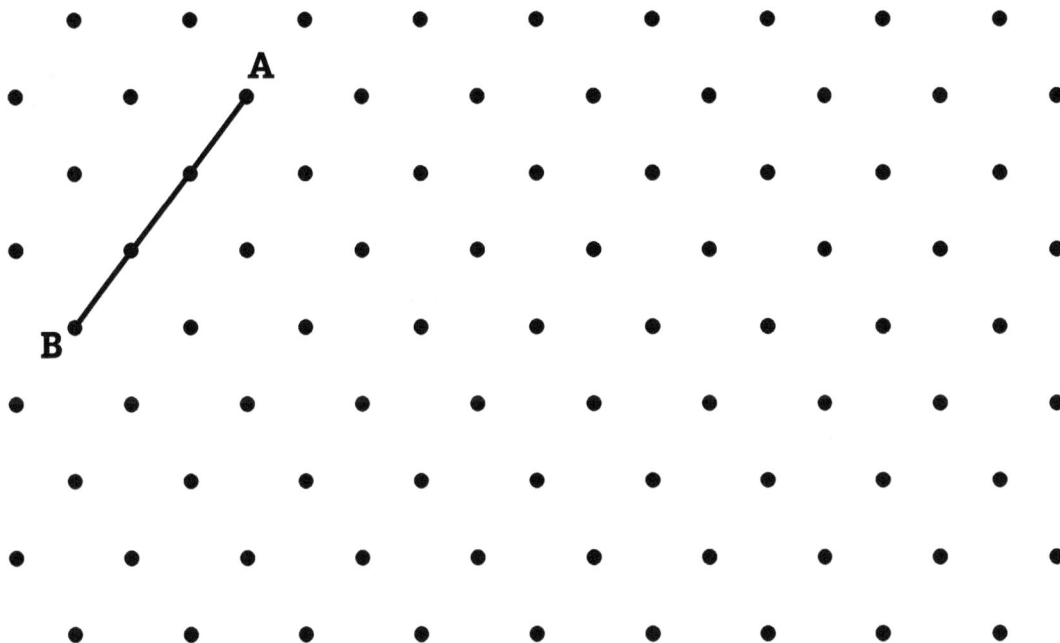

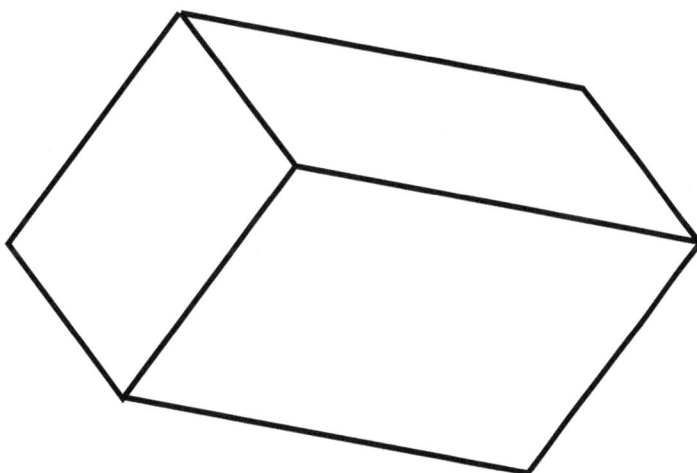

Geometry and Spatial Sense

10. Which of the figures below could be described using the following information?

    4 congruent sides; 2 lines of symmetry; opposite angles equal

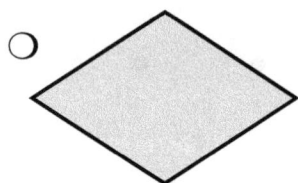

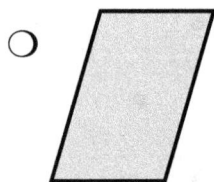

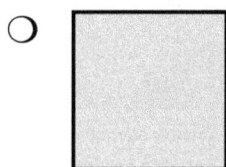

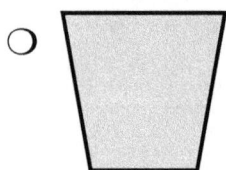

    ○

    ○

    ○

    ○

11. Manuel is sorting his geometric shapes into one of four categories: cones, prisms, pyramids, cylinders.

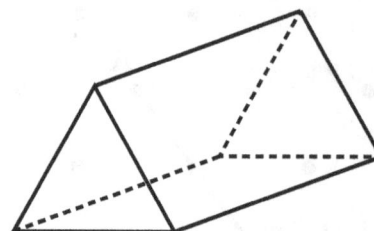

    Into which category should he place this figure?

    ○ cones

    ○ pyramids

    ○ prisms

    ○ cylinders

12. How many lines of symmetry does this letter have?

    D

    ○ none

    ○ one

    ○ three

    ○ two

SSG115   ISBN: 9781487704049 © On The Mark Press

Geometry and Spatial Sense

13. Use a ruler to determine the **perimeter** of the unshaded part of the rectangle below.

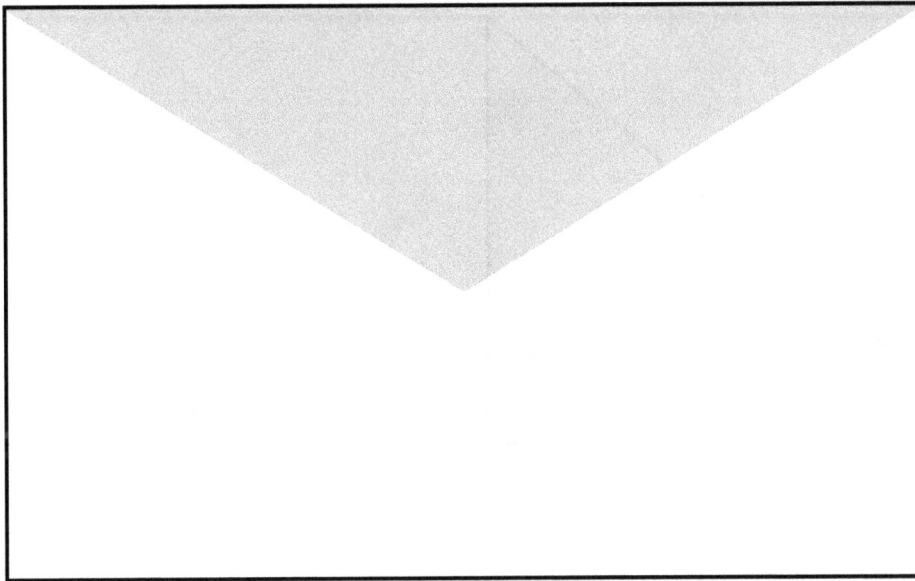

Show your work. Justify your answer.

The perimeter of the unshaded part of the rectangle is _____.

Geometry and Spatial Sense

14. Look at the triangle below and draw its reflection. Use a ruler to draw the sides and a protractor to measure and draw the angles.

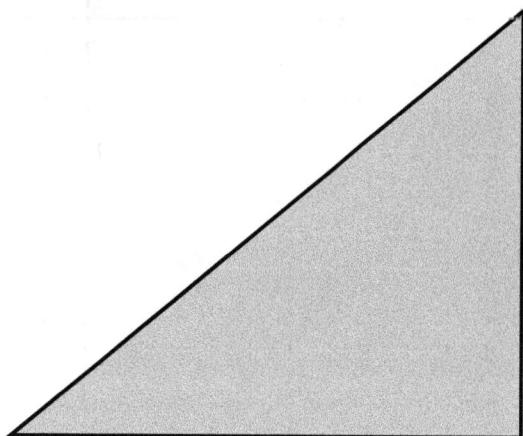

15. Use a ruler and draw the lines of symmetry in the following shapes.

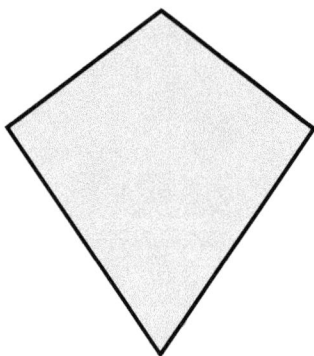

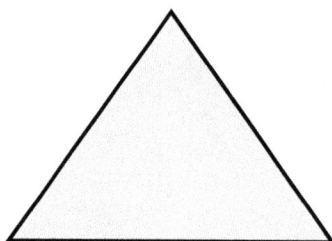

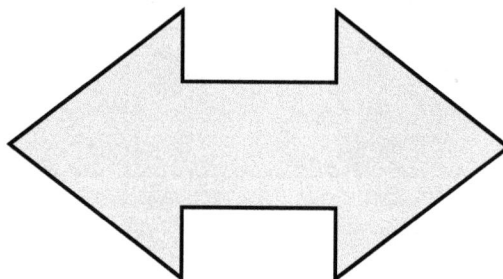

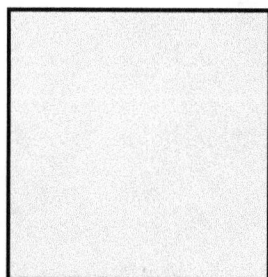

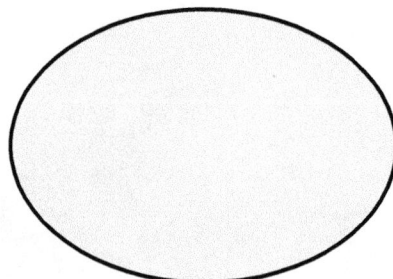

SSG115   ISBN: 9781487704049 © On The Mark Press

Management and Probability

1. Which of the following is the definition for the math term "range" when it is used to describe data?

   ○ found by dividing the sum of numbers by the number of numbers in a set

   ○ the difference between the highest and lowest values in a set of numbers

   ○ the middle number in a set of numbers arranged in order

   ○ the value that occurs most often in a set of data

2. Which of the following represents the probability that an event is not likely to occur?

   ○ 1

   ○ 0.75

   ○ 0

   ○ 0.12

3. This circle graph below shows what percentage of passengers get off at each stop on the bus route.

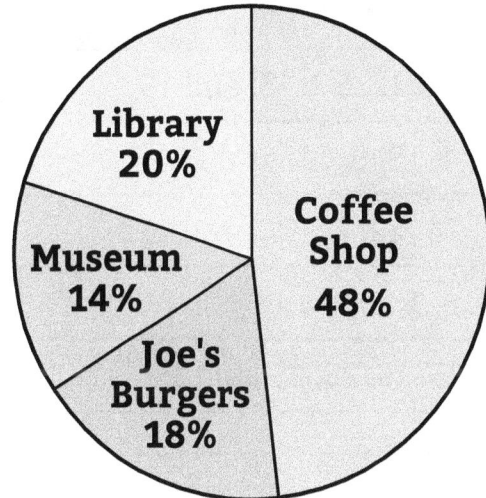

   If there are 65 passengers on the bus, how many people get off at the library?

   ○ 13

   ○ 14

   ○ 20

   ○ 18

Management and Probability

4. Complete the following chart. Determine the mode, median, mean and range of the sets of data to fill in the missing information.

Set of Data	Mode	Median	Mean	Range
a. 30, 12, 30, 13, and 19				
b. 4, 18, 6, 18, and 17				
c. 24, 22, 22, 20, and 17				
d. 6, 1, 6, 6, and 3				
e. 15, 15, 25, 15, and 15				

Show your work.

SSG115 ISBN: 9781487704049 © On The Mark Press

Management and Probability

5. There are numerous sources of fat in the Canadian diet. Make a bar graph to display the following data from Health Canada. Title and label your graph to reflect the information in the chart below.

<u>Sources of Fat in the Average Canadian Diet</u>

MEAT & ALTERNATIVES 26%

MILK PRODUCTS 17%

VEGETABLES & FRUIT. 2%

GRAIN PRODUCTS 13%

OTHER FOODS 42%

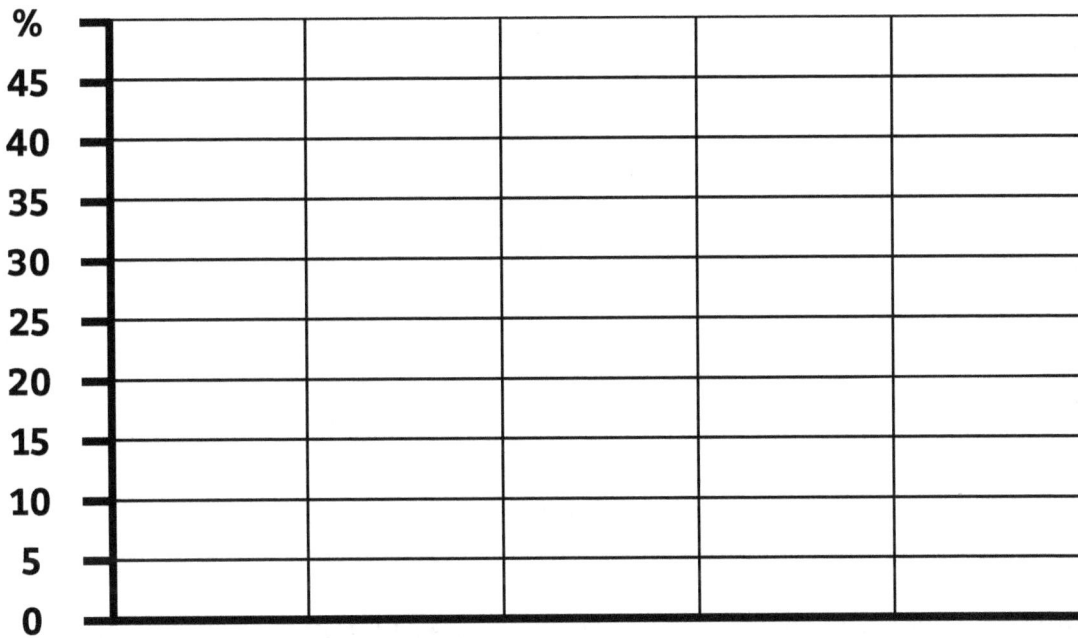

What food sources have the highest fat content? _____

Which food sources have the least fat content? _____

Management and Probability

6. In the box below, design a pie chart to best illustrate the following data. Two hundred free range chickens have produced the following egg records in one week:

25% LARGE EGGS	**12 % DOUBLE YOLK EGGS**
25% SMALL EGGS	**25 % MEDIUM EGGS**
12 % EXTRA LARGE EGGS	**1% CRACKED OR BROKEN EGGS**

Use a compass to create the pie chart.

This pie chart represents the farmer's total income of $400. After deducting $75 for expenses and loss of eggs, how much does each category on the pie chart generate towards the farmer's net income?

a. Large Eggs = _____   **d. Extra Large Eggs** = _____

b. Small Eggs = _____   **e. Double Yolk Eggs** = _____

c. Medium Eggs = _____

SSG115   ISBN: 9781487704049 © On The Mark Press

Management and Probability

7. Ross has a box of 40 blocks: 30 of the blocks are purple; 10 of the blocks are green.

   If Ross reaches into the box without looking and takes out 8 blocks, how many blocks can Ross expect to be purple?

   ○ 6 blocks

   ○ 4 blocks

   ○ 3 blocks

   ○ 8 blocks

8. Mohammed's family is going to paint their house. They have narrowed the choices to 4 colours but they can't decide which one they like best. They want to choose the colour randomly.

   Which of the following methods would be best for them to use?

   ○ toss a six-sided number dice with the numbers 1 – 6 on the faces

   ○ flip a coin

   ○ spin a spinner with 4 equal size sections labelled with the 4 colours

   ○ pick 1 card from 10 cards with 1 of the 4 colours written on each card

9. Keisha and Amy are keeping track of how many points they each score in the next 5 basketball games. They are recording their points in the table below.

Keisha's Points	15	18	9	6	15
Amy's Points	16	10	8	10	13

   According to the data, Keisha's mean of the points she scored is:

   ○ lower than Amy's mean of the points she scored

   ○ the same as Amy's range of the points she scored

   ○ higher than Amy's mean of the points she scored

   ○ the same as Amy's mode of the points she scored

Management and Probability

10. Look at the graph below. Carefully examine the monthly temperatures for Toronto.

Temperature Graph for Toronto

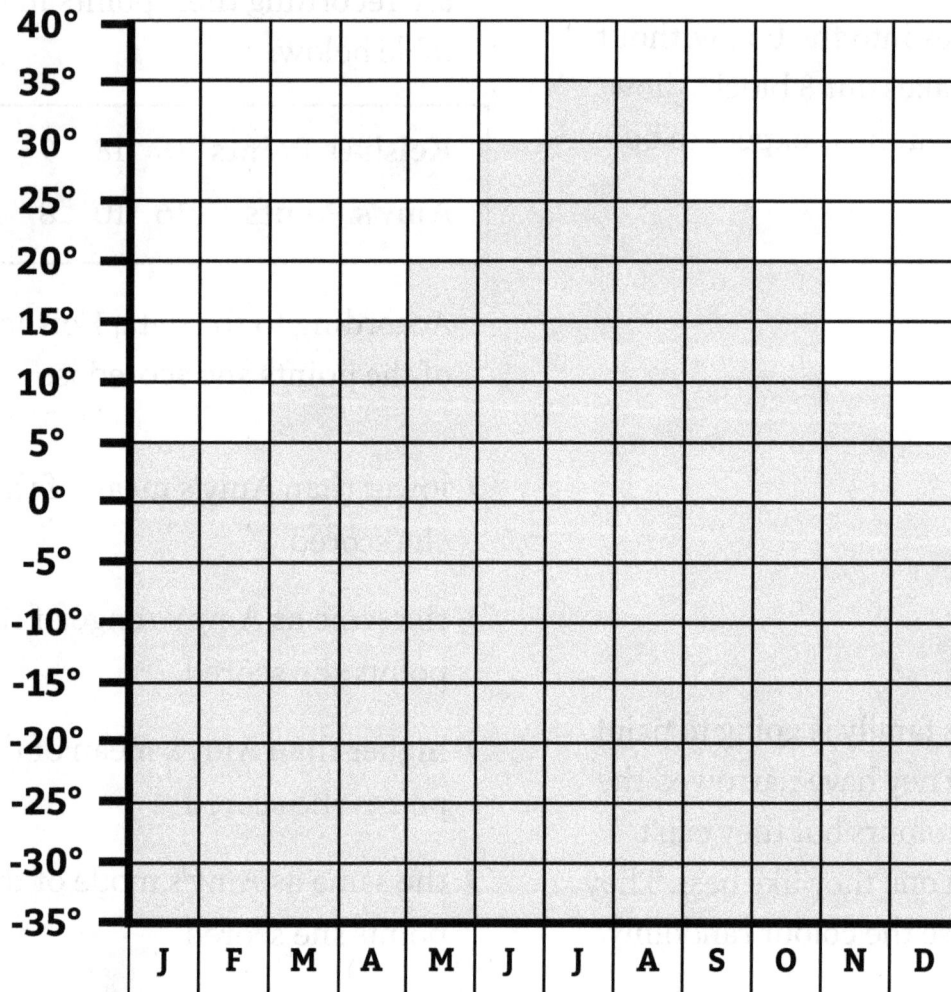

40°
35°
30°
25°
20°
15°
10°
5°
0°
-5°
-10°
-15°
-20°
-25°
-30°
-35°

J F M A M J J A S O N D

a. What is the mean of temperatures in Toronto? _____

b. What is the mode of temperatures in Toronto? _____

c. What is the range of temperatures in Toronto? _____

SSG115   ISBN: 9781487704049 © On The Mark Press

Management and Probability

11. The spinners below show 4 different games of chance. If the spinner lands on black you win $10. If it lands on white you don't win anything.

Spinner 1	Spinner 2	Spinner 3	Spinner 4
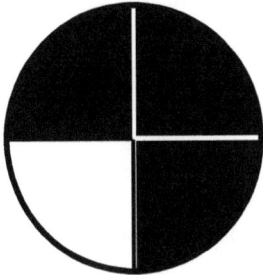		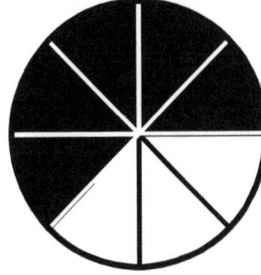	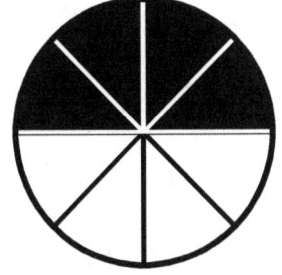

Show your work.

a. Which of the spinners gives you the best chance to win $10? _____

b. Which two spinners give you an equal chance of winning $10 or not winning anything? _____

c. Which spinner gives you the least chance of winning $10? _____

Mixed Math Skills

1. Which of the following set of temperatures is in order from lowest temperature to highest temperature?

   ○ 0°, 31°, -34°

   ○ 5°, 28°, -20°

   ○ -24°, -5°, 10°

   ○ 4°, -6°, 15°

2. Liz earns $600 per week. She started a savings account and puts in 15% of her earnings. How much money goes into Liz's savings account each week?

   ○ $60

   ○ $90

   ○ $50

   ○ $120

3. Which of the following is the sum of the angles of a rhombus?

   ○ 180°

   ○ 90°

   ○ 360°

   ○ 145°

4. Which of the following statements is true about the figures below?

   Figure A   **Figure B**

      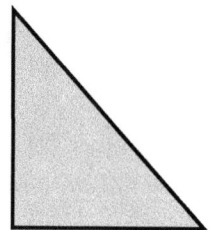

   ○ Figure A is an acute triangle.

   ○ Figure A is a right triangle.

   ○ Figure B is a scalene triangle.

   ○ Figure A and B are both acute triangles.

SSG115   ISBN: 9781487704049 © On The Mark Press

Mixed Math Skills

5. Look at the number pattern below.

75, 63, 51, 39, 27, . . .

Which rule describes how to find the next term in the pattern?

- ○ subtract 8 from the previous term

- ○ divide the previous term by 3

- ○ divide the previous term by 5

- ○ subtract 12 from the previous term

6. Listed below are the lengths of 5 pieces of rope:

   length 1 = 12.6 m

   length 2 = 9.75 m

   length 3 = 3.56 m

   length 4 = 26.08 m

   length 5 = 31.5 m

What is the total length of these 5 pieces of rope?

- ○ 83.49 m

- ○ 63.49 m

- ○ 53.29 m

- ○ 73.49 m

7. The graph below shows the amount of money Louis earned selling tickets for the school carnival over 4 days.

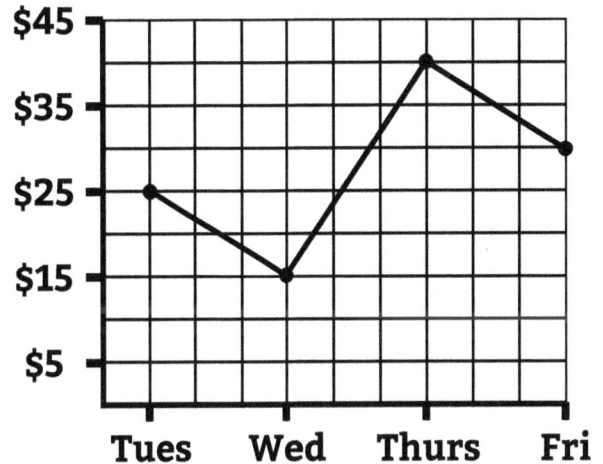

What is the range between the smallest and largest amount of money Louis earned over the 4 days?

- ○ $15

- ○ $25

- ○ $30

- ○ $10

Mixed Math Skills

8. A series of numbers is shown below:

    51, 82, 86, 55, 32, 81, 58, 34, 57, 84

    Which of the following stem and leaf plots represents these numbers?

    ○
Stem	Leaf
3	8, 9
5	1, 2, 7
8	1, 2, 4, 6

    ○
Stem	Leaf
3	2, 4
5	5, 8, 9
8	1, 2, 4, 6

    ○
Stem	Leaf
3	2, 4
5	1, 5, 7, 8
8	1, 2, 4, 6

    ○
Stem	Leaf
3	2, 4
5	1, 2, 7, 8
8	1, 2, 4, 6

9. Martin has $24.00 to spend at the mall. He used 0.7 of the $24.00 to buy a pair of jeans. How much did the jeans cost?

    ○ $2.40

    ○ $16.80

    ○ $ 10.80

    ○ $20.40

10. The cargo ship set sail from its home port at 3:20 a.m. It arrived at the port of San Paulo at 10:30 p.m. How long did it take the ship to get from its home port to San Paulo?

    ○ 19 hours and 10 minutes

    ○ 10 hours and 40 minutes

    ○ 12 hours and 50 minutes

    ○ 14 hours and 50 minutes

SSG115   ISBN: 9781487704049 © On The Mark Press

Mixed Math Skills

11. Fill in the missing sections of the chart below. The first problem is done for you.

	Input	Operation	Output	Number Sentence
A	(18, 7)	addition	25	18 + 7 = 25
B	(32, 4)		8	= 8
C	(5, 20)	multiplication		
D	(___, 6)	multiplication	42	
E	(24, 9)			24 – 9 = 15
F	(72, ___)	division	9	
G				16 × 6 = 96
H	(8, 8)		0	
I	(___, 62)	addition	76	
J	(21, 3)		7	

Mixed Math Skills

12. The area of a rectangle is 72 cm². Which of the following are possibly the dimensions of that rectangle?

○ length = 8 cm   width = 10 cm

○ length = 9 cm   width = 9 cm

○ length = 6 cm   width = 8 cm

○ length = 9 cm   width = 8 cm

13. The population of Bakersville is 4250. A new tech company is moving to town and the expected population growth is 20%. Which of the following numbers represents what the new population will be?

○ 4270

○ 5100

○ 6100

○ 5250

14. When Roshana and her Dad went to the farmer's market, they bought $18.84 worth of fruits and vegetables. Roshana's dad paid with a $20 bill. Which of the following should he get back in change?

○ one dollar, 2 dimes, 1 nickel, 1 cent

○ one dollar, 2 quarters, 1 cent

○ 4 quarters, 1 dime, 1 nickel, 1 cent

○ one dollar, 1 dime 1 cent

15. Lenny's small boat can travel 20 kilometres per hour. If he travels nonstop for 4 days, how far will he go?

○ 820 kilometres

○ 1920 kilometres

○ 1200 kilometres

○ 1440 kilometres

SSG115   ISBN: 9781487704049 © On The Mark Press

Mixed Math Skills

16. The owner of the Snack Shack kept track of how many cups of coffee and how many cups of tea were sold over a 4–day period. The table below shows how many cups of coffee were sold each day and how many cups of tea were sold during 2 of the 4 days.

	Day 1	Day 2	Day 3	Day 4	Mean of Cups Sold
Cups of Coffee	68	74	82	76	
Cups of Tea	84		70		

To complete the table, first figure out the mean of the cups of coffee sold.

The mean for the cups of tea sold is 20 cups more than that of the coffee sold. Using this information what are two possible numbers for the cups of tea sold on Day 2 and Day 4?

Fill in the table with your answers.

Show your work. Justify your answers.

Mixed Math Skills

17. Look at the shapes shown below. What is the ratio of circles to squares?

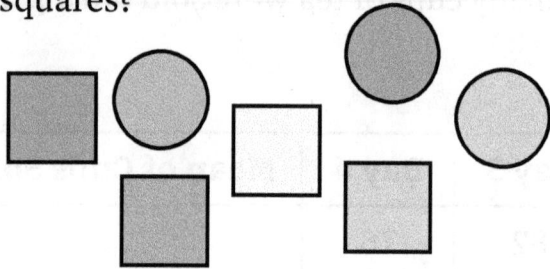

- ○ 3 to 4
- ○ $\frac{2}{6}$
- ○ 4 to 7
- ○ 3 to 7

18. How many degrees of the circle was taken away in the missing portion?

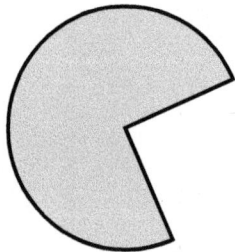

- ○ 45°
- ○ 15°
- ○ 90°
- ○ 25°

19. Liam bought 5 large orchids at the flower market for $80. To make an overall profit of $3, how much must he sell each orchid for?

- ○ $83.00
- ○ $16.60
- ○ $15.60
- ○ $15.00

20. The courthouse was built 22 decades ago. How many years ago was the courthouse built?

- ○ 22 years
- ○ 200 years
- ○ 220 years
- ○ 20 years

SSG115   ISBN: 9781487704049 © On The Mark Press

Mixed Math Skills

21. Emma and Ben were on a camping trip. They hiked 54 kilometres in 5 days.

    On the first day they hiked 10 kilometres. On the second day they hiked 8 kilometres. On the third day they hiked 15 kilometres. On the fourth day they hiked 12 kilometres.

    How many kilometres did they hike on the fifth day?

    ○ 11 kilometres

    ○ 9 kilometres

    ○ 13 kilometres

    ○ 6 kilometres

22. Lindsey's grandma was born in 1948. How old will she be in 2017?

    ○ 69 years old

    ○ 67 years old

    ○ 58 years old

    ○ 59 years old

23. Sanjay bought a bike on sale for $350. If 8% of that amount was sales tax, how much of the $350 was tax?

    ○ $80

    ○ $28

    ○ $35

    ○ $30

24. Which of the following represents how many degrees there are in a straight line?

    ○ 90°

    ○ 360°

    ○ 45°

    ○ 180°

Mixed Math Skills

1. There are two adjacent angles on a straight line. One of the angles is equal to 120°. What is the value of the second angle?

    ○ 145°

    ○ 60°

    ○ 90°

    ○ 35°

2. Determine the pattern rule for the number pattern below.

    360, 348, 336, 324, 312, . . .

    What will be the 9th term of this pattern?

    ○ 288

    ○ 264

    ○ 262

    ○ 276

3. Jason figured the area of a square to be 144 cm³. What was the length of one side of the square?

    ○ 14 cm

    ○ 10.4 cm

    ○ 12 cm

    ○ 12.2 cm

4. If Olivia rides her bike at an average speed of 5 km per hour, how many hours will it take her to travel 240 km?

    ○ 40 hours and 20 minutes

    ○ 41 hours

    ○ 48 hours and 20 minutes

    ○ 48 hours

SSG115   ISBN: 9781487704049 © On The Mark Press

Mixed Math Skills

5. Which fraction represents the shaded area in the figure below?

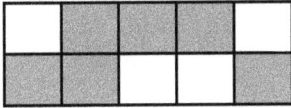

- ○ $\dfrac{3}{5}$
- ○ $\dfrac{6}{8}$
- ○ $\dfrac{5}{10}$
- ○ $\dfrac{4}{10}$

6. Which of the following equals the same amount as 1 cubic decimetre?

- ○ 2 litres
- ○ 2 kilograms
- ○ 1 litre
- ○ 1 gram

7. Mr. Sampson owns a small general store. He bought 150 bales of goods for $114 per bale. He will sell the goods in his store at $125 per bale.

How much profit will Mr. Sampson make on the 150 bales?

- ○ $1650 in profit
- ○ $1875 in profit
- ○ $17 100 in profit
- ○ $18 750 in profit

8. Melissa has to wait one more week for her vacation to start. How many hours does Melissa have to wait?

- ○ 48 hours
- ○ 168 hours
- ○ 72 hours
- ○ 268 hours

Mixed Math Skills

9. One of the ice sculptures at the Winter Carnival was made up of 375 blocks of ice. Some of the blocks were coloured to give a dramatic effect.

Out of the 375 blocks of ice:

$\frac{3}{15}$ were **RED** $\frac{4}{15}$ were **CLEAR** $\frac{1}{15}$ was **YELLOW**

$\frac{2}{15}$ were **GREEN** $\frac{5}{15}$ were **BLUE**

How many blocks of each colour were used in the ice sculpture?

Show your work.

_____ blocks were red.   _____ blocks were blue.

_____ blocks were green.   _____ blocks were yellow.

_____ blocks were clear.

SSG115   ISBN: 9781487704049 © On The Mark Press

Mixed Math Skills

10. Members of the Garden Club want to beautify the school grounds by planting bulbs that will bloom in the spring. They planted tulip and daffodil bulbs.

The chart below shows how many bulbs the members planted.

Tulip Bulbs Planted	Daffodil Bulbs Planted
Kelley: 72	Riley: 85
Max: ?	Emma: 134
Xavier: 122	Sasha: 162
Brittney: 146	Lisa: ?

The mean of the amount of tulip bulbs planted is 130. The mean of the amount of daffodil bulbs planted is 120.

Complete the chart by determining how many bulbs Max and Lisa planted.

Show your work.

Max planted _____ tulip bulbs.

Lisa planted _____ daffodil bulbs.

Mixed Math Skills

11. Look at the geometric shapes in each of the grids below. Follow the instructions given and draw the shape in its new position.

Rotate the shape clockwise 90°. Use the dot as your rotation point.

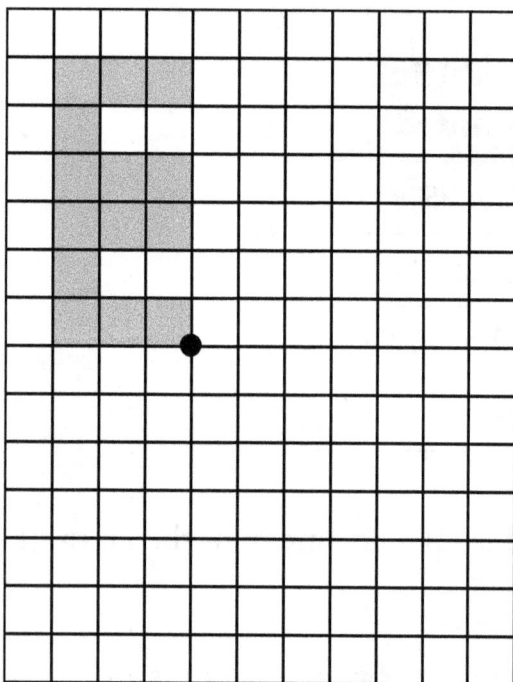

Flip the shape horizontally to create a reflection.

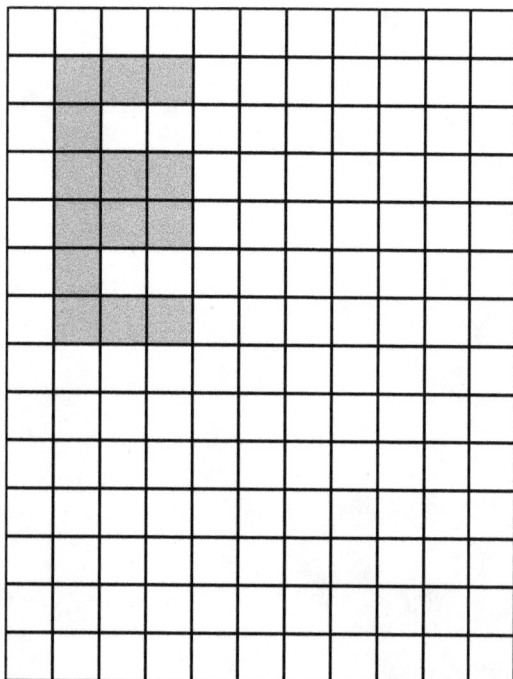

SSG115   ISBN: 9781487704049 © On The Mark Press

Mixed Math Skills

12. Jane has ridden her horse every day for 2 years. How many days has Jane ridden her horse?

   ○ 365 days

   ○ 650 days

   ○ 700 days

   ○ 730 days

13. Which of the following descriptions is an obtuse angle?

   ○ an angle that is 112°

   ○ an angle that is 68°

   ○ an angle that is 26°

   ○ an angle that is 216 °

14. Which of the following is a true statement about a triangular prism?

   ○ A triangular prism has 5 faces.

   ○ A triangular prism 6 vertices.

   ○ A triangular prism has 4 faces.

   ○ A triangular prism has 5 edges.

15. Mr. Ang has a container measuring 18 mm x 6 mm x 12 mm. What is the volume of Mr. Ang's container?

   ○ 1088 mm³

   ○ 1296 mm³

   ○ 288 mm³

   ○ 2880 mm²

16. What is the area of a triangle whose base measures 14.8 cm and whose height measures 6.8 cm?

   ○ 14.2 cm²

   ○ 100.63 cm²

   ○ 50.32 cm²

   ○ 144 cm²

17. How many degrees does the hour-hand of a clock move in 3 hours?

   ○ 180°

   ○ 90°

   ○ 45°

   ○ 15°

Mixed Math Skills

18. Andrea and Mitch are playing a game with a number cube. The cube is labelled 1 through 6. After rolling the cube, if it lands on a number that is a multiple of 2 the player gets 5 points. If the cube lands on a 5 the player gets 10 points. If the cube lands on a multiple of 3 the player gets 7 points.

   a. What is the probability of Andrea getting 5 points on her first roll?

   b. What is the probability of Mitch getting 10 points on his first roll?

   c. What is the probability of Andrea getting 7 points on her first roll?

Justify your thinking.

a. The probability of Andrea getting 5 points is _____.

b. The probability of Mitch getting 10 points is _____.

c. The probability of Andrea getting 7 points is _____.

SSG115   ISBN: 9781487704049 © On The Mark Press

Mixed Math Skills

19. Convert the following improper fractions to mixed numbers.

a. $\dfrac{9}{5}$ = _____

c. $\dfrac{7}{4}$ = _____

b. $\dfrac{8}{6}$ = _____

d. $\dfrac{8}{3}$ = _____

Show your work.

20. Reduce the following fractions to the lowest term.

a. $\dfrac{12}{16}$ = _____

c. $\dfrac{27}{81}$ = _____

b. $\dfrac{24}{40}$ = _____

d. $\dfrac{15}{18}$ = _____

Show your work.

Mixed Math Skills

1. The farm stand sells a bag of a dozen apples for $4.56. How much does one apple cost?

   ○ $1.25

   ○ $0.38

   ○ $0.35

   ○ $1.10

2. Which 2 equations are true if n = 4?

   EQUATION 1: $8 + n + 1 = 3$
   EQUATION 2: $8 - n - 1 = 3$
   EQUATION 3: $8 \times n - 1 = 3$
   EQUATION 4: $8 \div n + 1 = 3$

   ○ Equations 1 and 3

   ○ Equations 2 and 4

   ○ Equations 2 and 3

   ○ Equations 3 and 4

3. Jake earned $5.00 today. He put 15% of what he earned in his piggy bank.

   How much did Jake put in his piggy bank?

   ○ $1.00

   ○ $0.25

   ○ $1.75

   ○ $0.75

4. Which of the fractions below is the improper fraction for $5 \frac{3}{8}$?

   ○ $\frac{8}{8}$

   ○ $\frac{43}{8}$

   ○ $\frac{15}{8}$

   ○ $\frac{37}{8}$

SSG115 ISBN: 9781487704049 © On The Mark Press

Mixed Math Skills

5. The book club is helping to organize the books in the school library by category.

    The chart below shows the number of books in the following categories.

Type of Book	Quantity
Adventure	320
Mystery	168
Science Fiction	645
Art	122
Science	288
Biography	89

    Which of the following represents the range of the data shown?

    ○ 272

    ○ 556

    ○ 288

    ○ 320

6. Which of the following is the product of 63 and 17?

    ○ 1071

    ○ 46

    ○ 1270

    ○ 80

7. Look at the events listed below.

    EVENT 1: School will start on time on Monday.

    EVENT 2: The Prime Minister of Canada will give a speech at school on Monday.

    Which of the following represents the probability of these two events happening?

    ○ Event 1: Likely; Event 2: Likely

    ○ Event 1: Unlikely; Event 2: Impossible

    ○ Event 1: Likely; Event 2: Unlikely

    ○ Event 1: Impossible; Event 2: Unlikely

Mixed Math Skills

8. Starting at 8:00 a.m. every morning, a popular television station shows 20 minutes of commercials for every hour of programming. This lasts until 10:00 p.m. each day.

   How many minutes of programming is shown each day during this time period?

   How many minutes of programming is shown in one year during this time period?

Show your work. Explain your thinking.

There are _____ minutes of programming each day.

There are _____ minutes of programming each year.

SSG115   ISBN: 9781487704049 © On The Mark Press

Mixed Math Skills

9. Use a ruler to draw a rectangle and a triangle with the same area in the box below. Label the length and width of the rectangle and the base and height of the triangle.

Draw your rectangle and triangle here.

Mixed Math Skills

Mixed Math Skills

10. Fill in the stem and leaf plots to represent the two sets of numbers shown below.

a. 48, 49, 93, 77, 95, 44, 72, 98, 71, 79

Stem	Leaf

b. 66, 25, 11, 15, 22, 64, 13, 68, 63, 19

Stem	Leaf

SSG115   ISBN: 9781487704049 © On The Mark Press

Mixed Math Skills

11. The chart below shows how decimals, percentages, and fractions are related.

   Fill in the missing information to complete the chart.

	Decimal	Percent	Fraction
A	.4		
B			$\frac{4}{5}$
C	.25		
D			$\frac{8}{25}$
E		75%	
F		15%	
G	.65		
H			$\frac{7}{10}$

Mixed Math Skills

12. Look at the number pattern below. Determine its pattern rule.

    256, 128, 64, 32, 16, . . .

    Which of the following numbers is the next term in the pattern?

    ○ 4

    ○ 6

    ○ 8

    ○ 12

13. Rowena lives in a city with a population of seventy-six thousand twenty-five.

    Which number below represents the population of her city?

    ○ 76 025

    ○ 760 025

    ○ 7625

    ○ 76 250

14. Which of the following triangles has a perimeter of 12 cm and an area of 6 cm²?

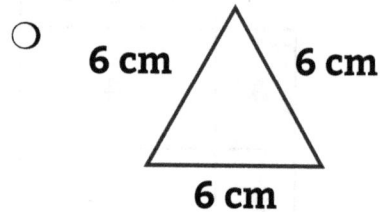

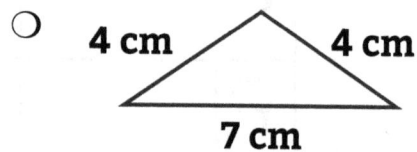

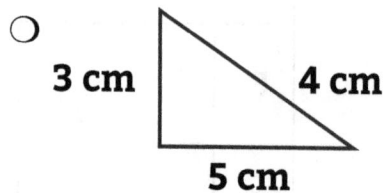

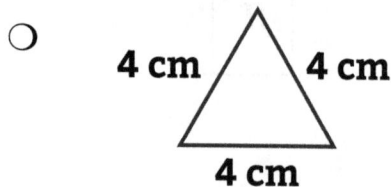

    ○

    6 cm   **6 cm**

    6 cm

    ○

    4 cm   **4 cm**

    7 cm

    ○

    3 cm   **4 cm**

    5 cm

    ○

    4 cm   **4 cm**

    4 cm

SSG115   ISBN: 9781487704049 © On The Mark Press

Mixed Math Skills

15. One section of the animal shelter has 48 dogs. There are 30 brown dogs.

    The rest of the dogs are black.

    Which of the following represents the ratio of black dogs to brown?

    ○ 30:48

    ○ 3:5

    ○ 5:3

    ○ 48:30

16. Lance and Mimi took a road trip for their vacation. They drove an average of 250 kilometres a day and were gone for 3 weeks.

    How many kilometres did they drive during their vacation?

    ○ 5500 km

    ○ 3500 km

    ○ 5250 km

    ○ 4250 km

17. Which of the following is equivalent to 3 ml of water?

    ○ 0.03 L

    ○ 0.3 L

    ○ 3 L

    ○ 0.003 L

18. Which of the following represents 119 960 when its rounded to the nearest then thousand?

    ○ 119 000

    ○ 120 000

    ○ 120 900

    ○ 119 900

Mixed Math Skills

1. Write the coordinates of the items shown on the grid.

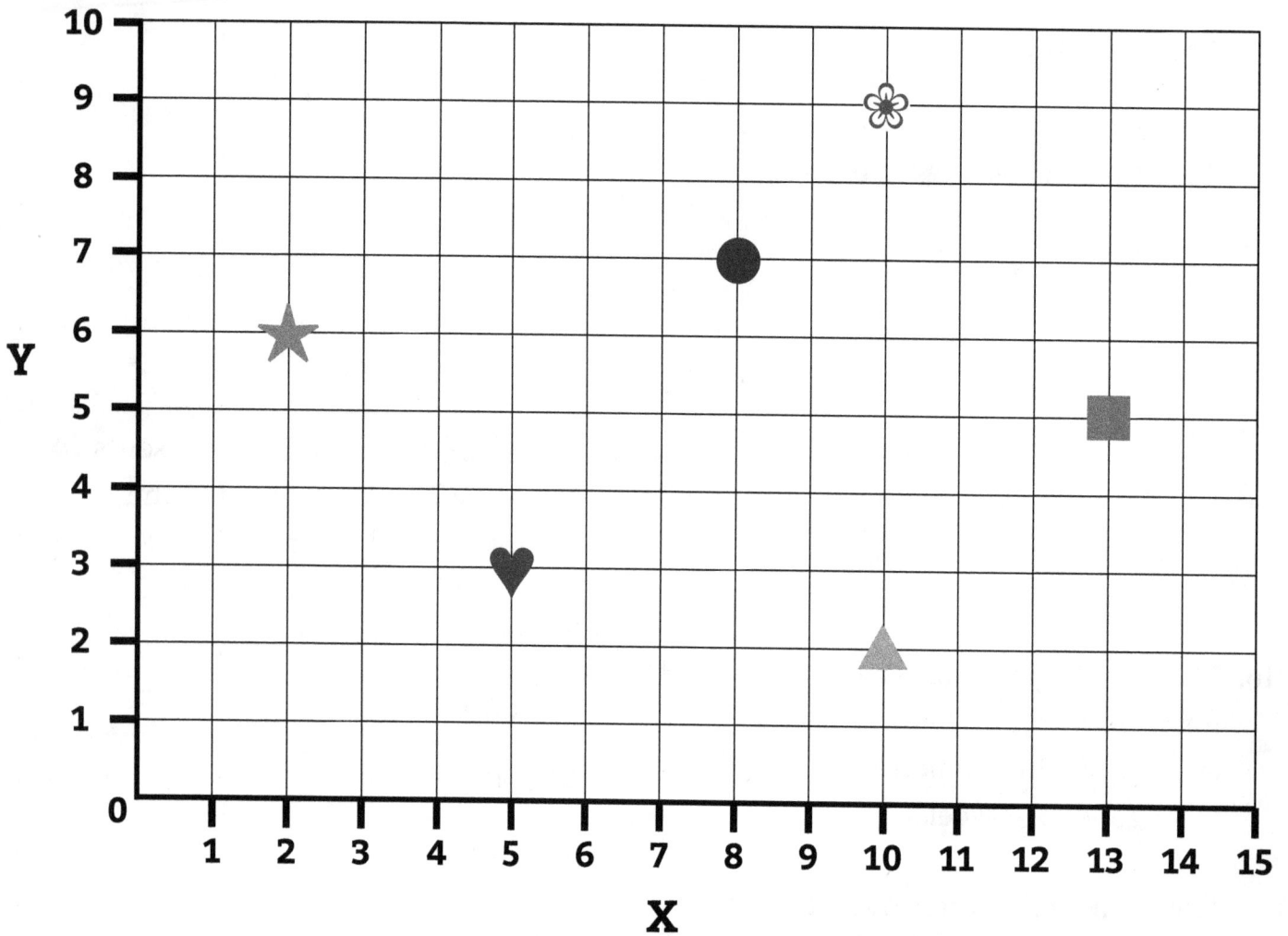

a. The heart: _____

b. The star: _____

c. The flower: _____

d. The square: _____

e. The triangle: _____

f. The circle: _____

SSG115   ISBN: 9781487704049 © On The Mark Press

Mixed Math Skills

2. Look at the shapes below. Some of the shapes are congruent and some of the shapes are similar. Use the letter code for each shape to identify which of the shapes are congruent and which are similar.

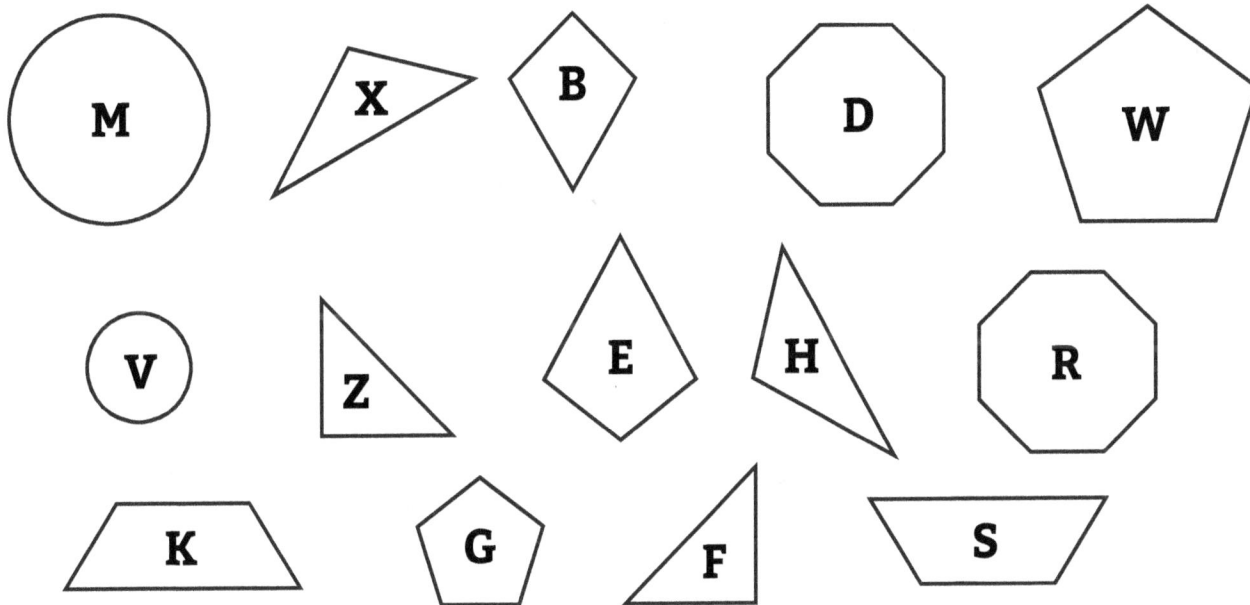

CONGRUENT	SIMILAR
_____	_____
_____	_____
_____	_____
_____	_____
_____	_____
_____	_____
_____	_____

Mixed Math Skills

3. The faces of Sean's number cube are labelled 1, 2, 3, 4, 5, and 6. Sean rolls the cube 36 times.

    How many times should Sean expect to roll a 4?

    ○ 4 times

    ○ 12 times

    ○ 6 times

    ○ 8 times

4. Look at the equation below.

    $$y \div z = 8$$

    Which values of y and z do not make the equation true?

    ○ y = 72; z = 9

    ○ y = 42; z = 7

    ○ y = 48; z = 6

    ○ y = 24; z = 3

5. Which of the following angles appear to measure 130°?

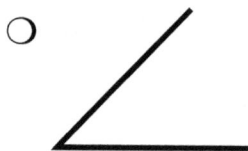

6. Which of the following numbers is expressed as 6 thousands, 9 millions, 4 ones and 8 tens?

    ○ 9 600 840

    ○ 9 006 084

    ○ 6 948

    ○ 9 648 000

SSG115   ISBN: 9781487704049 © On The Mark Press

Mixed Math Skills

7. Which of the following group of numbers are the prime numbers between 12 and 24?

   ○ 14, 17, 18. 20

   ○ 13, 15, 18, 21

   ○ 13, 17, 19, 23

   ○ 15, 16, 20, 22

8. How many lines of symmetry does the letter shown below have?

   V

   ○ none

   ○ two

   ○ three

   ○ one

9. Lora bought a pair of shoes, two sweaters and a purse at the mall. The cost of the items totalled $332.16. There is also a 15% tax on the items.

   How much was Lora's total including the tax?

   ○ $498.24

   ○ $381.98

   ○ $371.00

   ○ $382.16

10. Which 24-hour time indicates exactly 9:37 p.m.?

    ○ 2137 hours

    ○ 2037 hours

    ○ 2237 hours

    ○ 1937 hours

Mixed Math Skills

11. The local paper had a Halloween Candy Contest. Readers were asked to vote for their favorite candy. The chart below shows the results of that contest.

Type of Candy	Number of Votes
Mars bars	185
Caramilk	216
Crispy Crunch	234
Aero bars	198
Turtles	185
Smarties	269
Bounty	172
Coffee Crunch	265

a. What kind of candy got the most votes? _____

b. What is the mode of the data collected? _____

c. What kind of candy got the least votes? _____

d. What is the range of the data collected? _____

e. How many people voted in the contest? _____

f. How many more people voted for Coffee Crunch than Aero bars? _____

SSG115   ISBN: 9781487704049 © On The Mark Press

Mixed Math Skills

12. Mrs. Singh has a party planning business. She is looking online for the best place to get a good deal on assorted party favors. She has narrowed the search down to 3 websites.

Look at the information below and help Mrs. Singh make her decision.

WEBSITE 1: PARTY GOODIES	**WEBSITE 2:** SAM'S PARTY PLUS	**WEBSITE 3:** A & S PARTY SUPPLIES
3 dozen party favors for $7.29	5 dozen party favors for $9.99	2 dozen party favors for $5.39
Free Shipping	Shipping: $4.80	Free Shipping

Which website would give Mrs. Singh the best deal on party favors?

Show your work. Explain your thinking.

_____ **would give Mrs. Singh the best deal.**

Mixed Math Skills

13. The high school conducted a survey of 100 students to see how they got to school each day. The students were asked if they walked to school, drove to school or took the bus. The results are shown in the circle graph below.

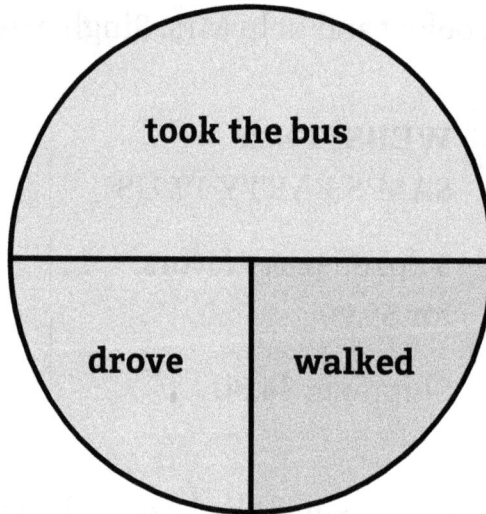

Express each answer to the following questions as a number, a percent, and a fraction.

	Number	Percent	Fraction
a. How many students took the bus?			
b. How many students walked?			
c. How many students drove or took the bus?			

SSG115   ISBN: 9781487704049 © On The Mark Press

Mixed Math Skills

14. What is the volume of a box with the following dimensions: height 8 cm, width 3 cm, length 6 cm?

   ○ 144 cm³

   ○ 24 cm³

   ○ 48 cm³

   ○ 120 cm³

15. Ramon is playing pool. The billiard balls shown below are on the pool table. When Ramon makes his next shot, what is the chance of him getting an odd numbered ball into a pocket?

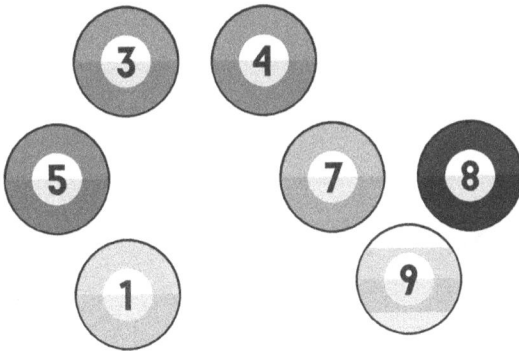

   ○ 1:2

   ○ 3:5

   ○ 5:7

   ○ 7:5

16. Which of the following is the radius of a circle with a diameter of 8 cm?

   ○ 2 cm

   ○ $\frac{1}{8}$ cm

   ○ 4 cm

   ○ 16 cm

17. Fresh picked fruit is available at the farmer's market. Peaches cost only 55¢ each if you buy a box of 2 dozen. How much does the box of peaches cost?

   ○ $6.60

   ○ $13.20

   ○ $12.50

   ○ $13.00

Mixed Math Skills

1. Mr. Nelson's art class is creating a mosaic design with coloured tiles. The design consists of 15 rows of tiles. The first row has 1 tile, the second row has 2 tiles, the third row has 3 tiles, the fourth row has 4 tiles and so on. This same pattern continues for all 15 rows.

   How many total tiles do the students need to complete the design?

   Show your work. Justify your thinking.

   The students will need _____ tiles.

SSG115   ISBN: 9781487704049 © On The Mark Press

Mixed Math Skills

2. Randy, Alex, Josh, and Ross all saved money to go to different sports camps. Randy spent ½ of his money on hockey camp. Alex spent 0.25 of his money on baseball camp. Josh spent 0.68 of his money on lacrosse camp. Ross spent ⅖ of his money on soccer camp.

Which boy has the highest percentage of his money left?

Show your work. Justify your thinking.

_____ **has the highest percentage of his money left.**

Mixed Math Skills

3. Draw the net for the three dimensional figure below. Show all vertices, faces, and edges.

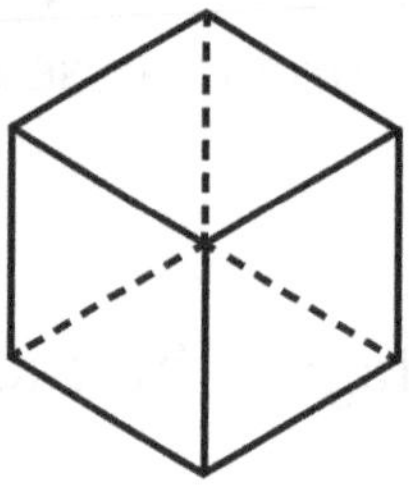

SSG115   ISBN: 9781487704049 © On The Mark Press

Mixed Math Skills

4. To save money, Sandy buys rice in the bulk and then shares it with her friends. She buys 10 kilograms of rice at a time. She gives each friend 500 grams of rice and keeps 500 grams for herself.

    How many friends does Sandy share her rice with?

    Show your work.

    Sandy shares her rice with _____ friends.

Mixed Math Skills

5. How many edges are there in a rectangular prism?

    ○ 12 edges

    ○ 15 edges

    ○ 9 edges

    ○ 10 edges

6. Which of the numbers below complete the following number sentence?

    (121 ÷ 11) × 3 + 5 = _____

    ○ 27

    ○ 38

    ○ 65

    ○ 42

7. What is the quotient of 14.007 by 7?

    ○ 2.1

    ○ 2.01

    ○ 2.010

    ○ 2.001

8. If you added one decimetre to 24 centimetres how many centimetres would you have?

    ○ 28 centimetres

    ○ 32 centimetres

    ○ 34 centimetres

    ○ 25 centimetres

SSG115   ISBN: 9781487704049 © On The Mark Press

Mixed Math Skills

9. Look at the squares below. One side of the outer square measures 5 centimetres.

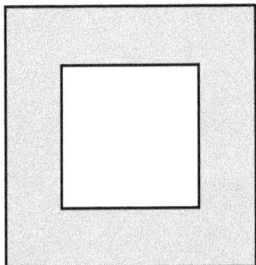

One side of the inner square measures 2.5 centimetres.

What is the area of the shaded portion?

○ 19 cm²

○ 18.75 cm²

○ 2.6 cm²

○ 25 cm²

10. Which of the following number sentences is true?

○ (56 + 33) + 13 = 56 + (33 + 13)

○ 45 + (16 + 110) = (4 + 5 + 110)

○ 2 + 29 = 25 + 2

○ 34 + (25 + 45) = (34 + 25) + 60

11. Misha and her family are on holiday in London. They have been there for 40 days. This represents 8/9 of the total days they will spend in London. How many days do Misha and her family have left before they go home?

○ 6 days

○ 10 days

○ 4 days

○ 5 days

12. How far will a train going 110 km an hour travel in 9 hours, if one hour is lost in stoppages?

○ 990 km

○ 880 km

○ 909 km

○ 808 km

Mixed Math Skills

13. A farmer wants to divide his farm amongst his four sons. The grid below shows the plan for his farm. Divide the grid into 4 equal portions. Each square equals 1 hectare. The side of each square is equal to 100 meters.

Farm Plan

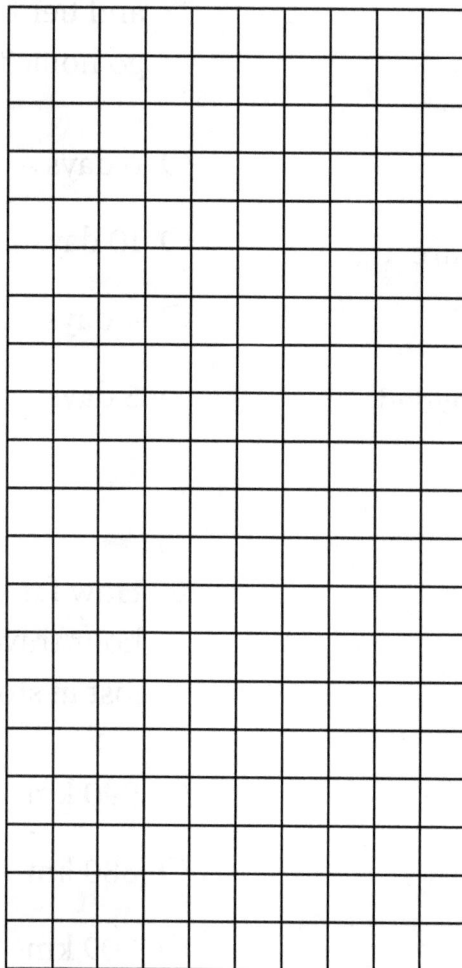

a. How many hectares will each son receive? _____

b. What is the perimeter of the whole farm? _____

c. What is the perimeter of each son's farm? _____

SSG115   ISBN: 9781487704049 © On The Mark Press

Mixed Math Skills

14. Fill in the precipitation graph below with the data in the chart for Vancouver, British Columbia.

Monthly Precipitation Data

Month	J	F	M	A	M	J	J	A	S	O	N	D
mm	214	161	151	90	69	65	39	44	83	172	198	243

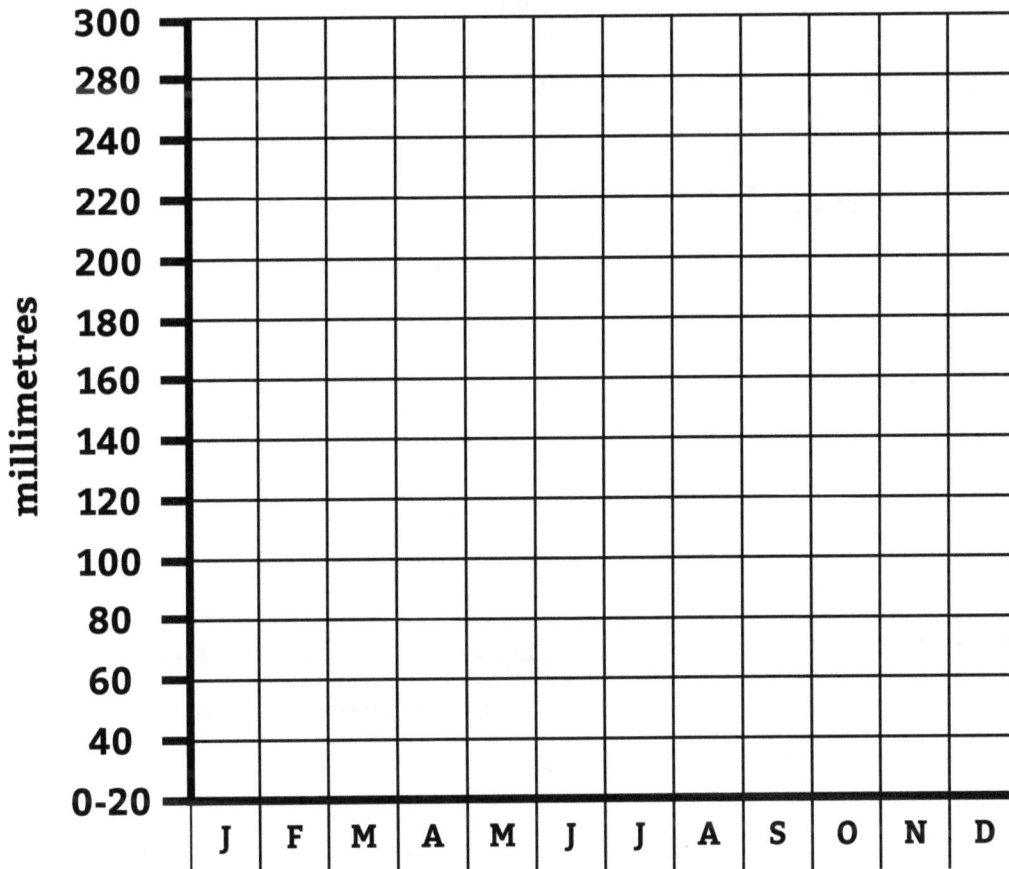

a. Which month gets the most precipitation? _____

b. What is the mean annual precipitation? _____

c. What is the range of the annual precipitation? _____

TEST #1: NUMBER SENSES AND NUMERATION

1. 12
2. 2.4
3. 0.89
4. $12,500
5. 6.93
6. $840
7. 50%
8. eighty thousand one hundred eighty-five
9. 10
10. $(800 \times 1\ 000) + (42 \times 1\ 000) + 71$
11. $850
12. 35:50, $^{35}/_{50}$ or $^{7}/_{10}$, 70%, 30%
13. $^{4}/_{10}$
14. $23 180
15. 3 hours
16. 8 rows of 12
17. 3 slices
18. a. 1,270 km   b. 11.54 hrs.   c. 7:30p.m.

TEST #2: PATTERNING AND ALGEBRA

1. Start with 2 and multiply by 3 to find the next term.
2. 8
3. 6, 15, 9, 18
4. 7
5. 2
6. 15,5

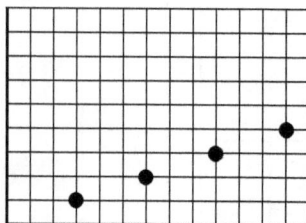

7. Day 5
8. Multiply by 3 and subtract 2.
9. ◄
10. a. 84   b. 924   c. 120   d. 3 432   e. 1 287   f. 6 435   g. 24 310
    h. 10   i. 5 005   j. 24 310
11. 10, 15, 5
12. 1 281
13. 520
14. 6
15. E

TEST #3: MEASUREMENT

1. 3 500
2. $12\ cm \times 25\ cm \times 18\ cm = 5\ 400\ cm^3$
3. Item 3
4. m^2
5. 70 km per hour
6. area: $48\ cm^2$ perimeter: 28 cm
7. 1500 tennis balls
8. 5:05 p.m.
9. $5\ cm \times 8\ cm = 40\ cm^2$
10. $4\ dm \times 5\ dm \times 8\ dm = 160\ dm^3$
11. 9 cm
12. 3000 millilitres
13. $50\ 000\ cm^2$
14. 11:10 a.m.
15. a. 1 $20 bill, 1 $10 bill, 1 $5 bill, 1 dollar coin, 1 quarter, 1 dime, 1 nickel   b. 1 $10 bill, 2 two dollar coins, 2 quarters, 1 dime, 1 nickel, 3 pennies
16. 340 minutes, 5 ½ hours

TEST #4: GEOMETRY AND SPATIAL SENSE

1. isosceles triangle
2. 6 lines of symmetry
3.
4.
5. 8 vertices
6. circle
7. obtuse
8. rectangle
9. image should match sample
10. rhombus
11. prisms
12. one
13. 38.5 cm
14.
15.

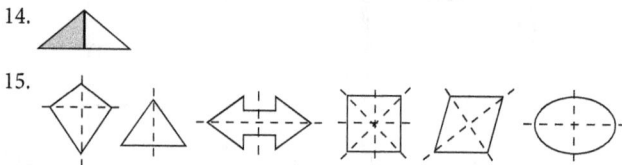

TEST #5: DATA MANAGEMENT AND PROBABILITY

1. the difference between the highest and lowest values in a set of numbers
2. 0.12
3. 13
4. a. 30, 19, 21, 18   b. 18, 17, 12.6, 14   c. 22, 22, 21, 7   d. 6, 6, 4, 4, 5
    e. 15, 15, 17, 10
5. other foods, vegetables and fruit
6. a. $81.25   b. $81.25   c. $81.25   d. $39.00   e. $39.00
7. 6 blocks
8. spin a spinner with 4 equal size sections labelled with the 4 colours
9. higher than Amy's mean of the points she scored
10. a. 17.92°   b. 5°   c. 30°
11. a. Spinner 1   b. Spinners 2 & 4   c. Spinner 3

SSG115   ISBN: 9781487704049 © On The Mark Press

TEST #6: MIXED MATH SKILLS

1. -24˚, -5˚, 10˚
2. $90
3. 360˚
4. Figure A is an acute triangle.
5. subtract 12 from the previous term
6. 83.49 m
7. $25
8.

Stem	Leaf
3	2, 4
5	1, 5, 7, 8
8	1, 2, 4, 6

9. $16.80
10. 19 hours and 10 minutes
11. b. division, $32 \div 4 = 8$ c. 100, $5 \times 20 = 100$ d. 7, $7 \times 6 = 42$
    e. subtraction, 15   f. 8 $72 \div 8 = 9$ g. (16, 6), multiplication, 96
    h. subtraction, $8 - 8 = 0$ i. 14, $62 + 14 = 76$
    j. division, $21 \div 3 = 7$
12. length = 9   width = 8 cm
13. 5100
14. 4 quarters, 1 dime, 1 nickel, 1 cent
15. 1920 kilometres
16. 75, 95, 110, 116 (answers will vary)
17. 3 to 4
18. 90˚
19. $16.60
20. 220 years
21. 9 kilometres
22. 69 years old
23. $28
24. 180˚

TEST #7: MIXED MATH SKILLS

1. 60˚
2. 264
3. 12 cm
4. 48 hours
5. ⅗
6. 1 litre
7. $1650 in profit
8. 168 hours
9. 75 red, 50 green, 100 clear, 125 blue, 25 yellow
10. Max 180, Lisa 99
11. a.   b.

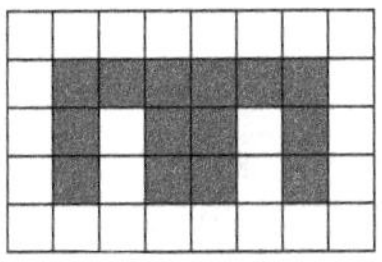

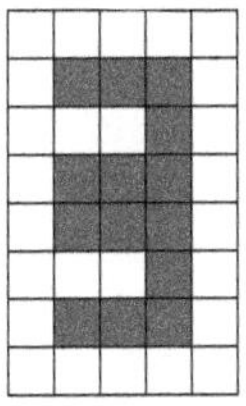

12. 730 days
13. an angle that is 112˚
14. A triangular prism has 4 faces.
15. 1296 mm³
16. 50.32 cm
17. 90˚
18. a. ½   b. ⅙   c. ²⁄₆
19. a. 1 ⅘   b. 1 ²⁄₆   c. 1 ¾   d. 2 ⅔
20. a. ¾   b. ⅗   c. ⅓   d. ⅚

TEST #8: MIXED MATH SKILLS

1. $0.38
2. Equations 2 and 4
3. $0.75
4. 43/8
5. 556
6. 1071
7. likely, unlikely
8. 560, 204 400
9. answers will vary
10. a. 4 – 4, 8, 9;   7 – 1, 2, 7, 9;   9 – 3, 5, 8
    b. 1 – 1, 3, 5, 9;   2 – 2, 5;   6 – 3, 4, 6, 8
11. a. 40%, ⁴⁄₁₀ or ⅖   b. 0.8, 80%   c. 25%, ¼   d. 0.32, 32%
    e. 0.75, ¾   f. .15, ³⁄₂₀   g. 65%, ¹³⁄₂₀   h. 0.7, 70%
12. 8
13. 76 025
14.

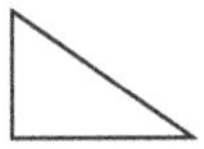

15. 3:5
16. 5250 km
17. 0.003 L
18. 120 000

TEST #9: MIXED MATH SKILLS

1. a. (5,3)   b. (2, 6)   c. (10, 9)   d. (13, 5)   e. (10, 2)   f. (8, 7)
2. Congruent: X & H, D & R, Z & F, K & S
   Similar: M & V, B & E, W & G
3. 6
4. y = 42 ; z=7
5.

6. 9 006 084
7. 13, 17, 19, 23
8. one
9. $381.98
10. 2137 hours
11. a. Smarties   b. 185   c. Bounty   d. 97   e. 1724   f. 67
12. Website 1
13. a. 50, 50%, ½   b. 25, 25%, ¼   c. 75, 75%, ¾
14. 144 cm³
15. 5:7
16. 4 cm
17. $13.20

TEST #10: MIXED MATH SKILLS

1. 120
2. Alex
3. Any of these possibilities:

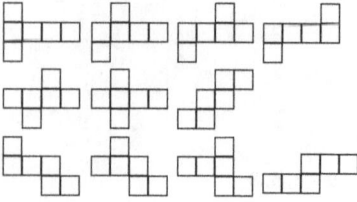

4. 19
5. 12 edges
6. 38
7. 2.001
8. 34 centimetres
9. 18.75 cm^2
10. $(56 + 33) + 13 = 56 + (33 + 13)$
11. 5 days
12. 880 km
13. a. 50 hectares   b. 6000 metres   c. 3000 metres
14. a. December   b. 127.4 mm   c. 204 mm

SSG115   ISBN: 9781487704049 © On The Mark Press